Programming Paradigms' Profound Potential

A Deep Dive Into Programming Languages And Their Underlying Principles

Alaric Hemingway

Contents

1 Introduction to Programming

Understanding a programming language begins with grasping its fundamental building blocks: variables, data types, and operators. Variables act as containers for storing data, like numbers, text, or more complex structures. Data types define the kind of information a variable can hold—integers for whole numbers, floating-point numbers for decimals, strings for text, and booleans for true/false values. Operators are symbols that perform operations on data, such as arithmetic operations (+, -, *, /), comparison operations (==, !=, <, >), and logical operations (&&, | |, !). Mastering these elements is the cornerstone of any successful programming journey. Think of them as the alphabet and grammar of any programming language; you can't write a sentence without knowing the letters and sentence structures.

Algorithms are sets of step-by-step instructions to solve a specific problem. They are the heart of any program. Learning to design and implement effective algorithms is crucial. Consider a simple task like sorting a list of numbers. You could approach it using various algorithms—bubble sort, merge sort, or quicksort, each with different efficiency characteristics. Understanding algorithm complexity, typically expressed using Big O notation, helps you choose the most appropriate algorithm for a given task, crucial for building efficient and scalable software. For example, a simple linear search algorithm is sufficient for small data sets but will become drastically slow when dealing with millions of data points, therefore choosing a better algorithm like binary search, which is far more efficient, becomes necessary.

Control structures dictate the flow of execution within a program. Conditional statements like if, else if, and else allow your program to make decisions based on conditions. Loops, including for and while loops, enable repetitive execution of code blocks. These structures are essential for creating programs that perform complex tasks dynamically and adaptively. Imagine creating a program to check if a number is prime. You would first use a loop to check divisibility by numbers less than the given number's square root. Then, you would use an if condition to check if any of these numbers perfectly divides the given number. This showcases how conditional statements and loops are essential for making decisions in programs.

Functions are reusable blocks of code that perform specific tasks. They enhance code organization, readability, and maintainability. Well-designed functions promote modularity, breaking down complex programs into smaller, manageable units. Consider building a calculator application. Instead of writing all the code for addition, subtraction, multiplication, and division in a single block, functions would be the perfect tool. Creating dedicated functions for each operation not only improves readability but also simplifies debugging and future modifications. Using functions can make large code bases much easier to handle.

Debugging is the process of identifying and fixing errors in your code. It's an integral part of programming, requiring patience and systematic problem-solving skills. Techniques include using a debugger (a tool that lets you step through your code line by line), printing intermediate values to the console for inspection, or utilizing testing frameworks to automate the verification of your code's correctness. Debugging is crucial, particularly when handling complex and intricate codes. If your program produces unexpected outputs, using debug tools allows you to observe your variable values and understand where they deviate from expectations, leading you to find the source of the problem.

Developing good programming practices from the beginning is crucial for long-term success. This involves adopting a structured approach to coding, adhering to style guidelines, writing clear and concise comments to explain your code, and performing thorough testing. Following these practices enhances code maintainability, allowing you to update, expand and debug your code more efficiently in the future. They also improve code readability, making it significantly easier for others to

understand your code and collaborate with you.
Finally, remember that learning to program is a continuous process. There will be challenges, errors, and moments of frustration. Persistence and a willingness to learn from mistakes are essential. Leverage online resources, participate in coding communities, and practice regularly. The more you code, the more confident and proficient you will become. Continuous learning and practice are fundamental aspects of achieving proficiency and mastery in programming. Continuous learning, seeking help from communities, and consistent practice are essential to mastering programming's many facets.

1.1 What is a Paradigm?

A paradigm, in the context of programming, isn't just a style or a trend; it's a fundamental model, a worldview that dictates how we conceptualize and solve problems using code. It's the underlying philosophy that shapes the structure, design, and execution of a program. Think of it as the lens through which you view the problem, influencing every decision from the initial design to the final implementation. This perspective is crucial because it directly impacts the way we approach software development—from the tools we utilize to the strategies we employ. Choosing the right paradigm is akin to selecting the perfect tool for a job. A hammer won't be effective when you need a screwdriver, and similarly, applying an object-oriented approach to a problem best suited for functional programming will lead to inefficiency and potentially, flawed code.

Understanding paradigms isn't merely about learning syntax; it's about grasping the core principles and the trade-offs involved in each approach. It's about choosing the right paradigm to effectively address the specific challenges inherent in a problem and optimizing for factors like efficiency, readability, and maintainability. For example, dealing with complex data transformations may benefit from the elegant and concise nature of functional programming, while building user interfaces might favor the intuitive structure of object-oriented programming. Mastering these different approaches requires recognizing their strengths and limitations and applying them strategically based on the context of the program being developed.

Different paradigms offer distinct ways to represent data and control the flow of execution. In imperative programming, for instance, you explicitly instruct the computer on each step it should take, sequentially modifying the program's state. This is like giving a detailed set of instructions to a robot, specifying every movement and action. Object-oriented programming, on the other hand, utilizes objects and classes, promoting modularity and reusability. This is akin to building a machine from pre-fabricated parts, assembling them according to their defined functionalities. Functional programming, meanwhile, focuses on mathematical functions and immutable data, minimizing side effects and simplifying the process of reasoning about code. This is like assembling a complex structure using standardized components where each component's operation is independent of others.

Furthermore, logic programming presents yet another approach, focusing on logical statements and their relationships to derive solutions. This is like using a sophisticated inference engine to solve puzzles based on established rules and facts. Each approach carries its own set of benefits and disadvantages. Imperative programming can be explicit and intuitive but often struggles with complexity. Object-oriented programming enhances modularity but may face difficulties with concurrency. Functional programming promotes purity but could demand more advanced comprehension. Logic programming boasts declarative power but might not be suited for all problem domains. The key takeaway is that no single paradigm is universally superior; the optimal choice depends entirely on the specific problem and project requirements.

Consider the task of building a simple calculator. While you could certainly implement it using an imperative approach, meticulously detailing every step of addition, subtraction, multiplication, and division, this might become cumbersome as complexity increases. An object-oriented approach, where you create objects representing numbers and operations, might prove more organized. You could design classes for different types of calculations, making the code more modular and easier to extend. Similarly, a functional approach, by creating functions for each operation, could promote code reusability and eliminate side effects. Selecting the best paradigm is not arbitrary but a decision driven by the nature of the problem at hand and the long-term considerations of scalability and maintainability. Understanding this critical aspect of design is paramount to becoming an effective programmer. This crucial choice lays the foundation for a robust, efficient, and well-structured program. Therefore, a deep understanding of programming paradigms is vital for any aspiring software developer. The correct paradigm simplifies the development process and results in elegant and efficient solutions. The choice of a paradigm profoundly influences the quality, efficiency, and maintainability of the software.

1.2 Paradigm History

The earliest forms of programming, predating the concept of a "programming language" as we know it, involved directly manipulating hardware. Think of punch cards and wiring panels – these were the initial interfaces, dictating the flow of electricity to achieve specific computations. These rudimentary methods, while incredibly primitive by today's standards, represent the genesis of programmatic thought; a direct, albeit painstaking, interaction with the machine's fundamental workings. The limitations were immense, requiring meticulous planning and a deep understanding of the underlying hardware architecture. It was a time when the programmer was intrinsically tied to the physical machine, a far cry from the abstract environments of modern software development. This direct hardware manipulation, however, laid the groundwork for all subsequent programming paradigms, establishing the core principle of instructing a machine to perform specific tasks.

The emergence of assembly languages marked a significant step forward. Instead of manipulating binary code directly, programmers could now use mnemonics, symbolic representations of machine instructions. This abstraction, while seemingly minor, drastically improved the programming process, making it less error-prone and more manageable. The transition to symbolic representation made programming less about intricate hardware details and more about logical problem-solving. However, assembly languages remained highly machine-specific, meaning code written for one machine wouldn't work on another. Programmers still needed intimate knowledge of the target hardware, restricting code portability and hindering the development of large-scale software projects. Despite its limitations, the introduction of symbolic representation was a pivotal moment in programming paradigm history, representing the first shift away from raw machine interaction.

The invention of high-level languages brought about a profound transformation. FORTRAN, one of the earliest examples, permitted programmers to express algorithms in a manner more akin to mathematical notation. This represented a significant leap towards abstraction, moving away from the machine-specific nature of assembly languages. The ability to express computations in a more human-readable format facilitated the development of larger and more complex software systems. The shift toward higher-level abstraction not only increased programmer productivity but also fostered collaboration and knowledge sharing amongst developers. It was in this era that the concept of a "programming paradigm" started to truly take shape, although the full implications weren't immediately recognized.

The development of structured programming further refined the way programmers approached software development. Instead of using unstructured GOTO statements to control program flow (a common practice with earlier languages), structured programming emphasized the use of control structures like loops and conditional statements. This methodology led to more readable, maintainable, and less error-prone code. The focus shifted towards modularity and breaking down complex programs into smaller, more manageable units. This emphasis on organization and structure directly influenced the design and implementation of later programming paradigms, highlighting the enduring impact of structured programming on the software development landscape. It laid the foundation for the object-oriented approach, emphasizing code reusability and maintainability—aspects that remain central to software engineering today.

Object-oriented programming (OOP), introduced

decades later, presented a paradigm shift in its own right. It introduced the concepts of classes, objects, inheritance, and polymorphism, providing a more intuitive and powerful way to model real-world problems. OOP's emphasis on data encapsulation and abstraction resulted in more robust and maintainable software. The ability to reuse code through inheritance significantly improved programmer productivity and reduced development time. OOP's influence on software development has been profound and far-reaching, transforming how developers approach software design and shaping the landscape of modern software systems. Many widely-used programming languages today heavily rely upon OOP principles, showcasing its enduring impact.

The rise of functional programming offered yet another perspective on software development. This paradigm emphasizes the use of pure functions and immutability, promoting cleaner code that is easier to test, debug, and parallelize. Functional programming's focus on mathematical functions contrasts sharply with the state-changing nature of imperative programming. The ability to reason about code more precisely, thanks to the absence of side effects, makes functional programming highly suitable for complex systems requiring high reliability. Its increasing adoption in many areas underscores its potential and the growing recognition of its advantages in specific contexts.

The evolution of programming paradigms hasn't been a linear progression, with each new approach replacing its predecessors. Instead, different paradigms often coexist and complement each other, with programmers choosing the most suitable approach depending on the specific problem at hand. This ongoing evolution showcases the dynamic nature of software development and its continuous adaptation to evolving technological landscapes and changing programmer needs. The history of programming paradigms illustrates a constant pursuit for greater abstraction, modularity, and efficiency in solving increasingly complex problems.

2 Imperative Programming

Imperative programming, at its core, directs the computer through a sequence of explicit commands. Think of it like a detailed recipe: each step must be followed precisely, in order, to achieve the final result. This sequential nature is its defining characteristic, a fundamental difference from paradigms that prioritize declarations or logical relationships. The programmer explicitly defines every action the computer takes, leaving little room for implicit behavior or inference. This direct control, while offering clarity in simple programs, can quickly lead to complex and difficult-to-manage codebases as the scale of the project increases.
Procedural programming, a prominent sub-category of the imperative approach, organizes code into procedures or functions. These functions, performing specific tasks, improve code reusability and modularity. Imagine building a house: you'd likely have separate procedures for laying the foundation, framing the walls, and installing the roof. Each procedure focuses on a

specific aspect of the overall project, making the construction process more manageable and less prone to errors. However, the procedural approach's focus remains on the steps, rather than the data itself—a pivotal distinction to consider when comparing it to more modern, data-centric approaches.

Structured programming, a refinement of the procedural approach, emphasizes the use of control structures like loops and conditional statements to organize the flow of execution. These structures, like if-else statements and for or while loops, allow for more flexible and efficient program design. Think of them as traffic signals guiding the execution path; they enable the program to make decisions and repeat actions, adapting its behavior based on input or changing conditions. Careful structuring is key to creating manageable and understandable imperative programs, even within a procedural framework. Without such structure, the code risks becoming a tangled mess of instructions, difficult to maintain or debug.

Control structures are the backbone of imperative programming's ability to manipulate the program's state. They dictate which parts of the code are executed, and in what order, allowing for conditional logic and iterative processes. This manipulation of the program's state through assignment and modification of variables is central to the imperative paradigm's operational mechanisms. Statements like x = x + 1 explicitly change the value of the variable x, and this explicit control over state is a hallmark of imperative approaches. Mastering these constructs is crucial for any programmer working with imperative languages because they govern the program's behavior directly. Inefficient or poorly designed control structures are a common source of errors in imperative programs.

Beyond simple control structures, more sophisticated techniques are needed to manage complexity in larger programs. Modular design breaks down large programs into smaller, independent modules, promoting maintainability and reducing redundancy. Each module, ideally with a well-defined interface and a specific functionality, simplifies the understanding and modification of the code. This approach mirrors the compartmentalization seen in engineering projects: separating the design of the electrical system from the mechanical system makes each task more manageable and less prone to errors. Similarly, a modular approach in imperative programming allows multiple programmers to work concurrently on separate parts of the project.

Another crucial aspect of large-scale imperative programming involves managing data structures efficiently. Arrays, linked lists, stacks, and queues are all examples of data structures that can dramatically impact a program's performance. The choice of a suitable data structure often depends on the specific application and its performance requirements. Understanding how to choose and implement appropriate data structures is vital in writing efficient and performant imperative code. Failure to do so can lead to significant performance bottlenecks, especially with large datasets. An improper choice of data structure, for example, could lead to an algorithm with a complexity that scales poorly with data size.

Error handling in imperative programs is an essential aspect of building robust software. Imperative languages often provide mechanisms such as exception handling and error codes to deal with unexpected situations. A well-designed error-handling strategy will improve the program's stability and reliability. Imagine a program processing financial data: a failure to handle potential errors, like invalid input or network outages, could lead to inaccurate results or system crashes. Thus, a robust error handling plan is crucial, including mechanisms to catch, log, and gracefully manage those situations.

While imperative programming offers direct control and clarity in smaller projects, scaling to large, complex systems requires significant discipline and careful planning. The sequential nature of execution can make concurrent programming more challenging, and issues like race conditions and deadlocks necessitate careful synchronization mechanisms. These are complex issues requiring in-depth knowledge of concurrent programming techniques—a topic which is explored in detail later in this book. Understanding the limitations and strengths of the imperative paradigm is therefore key to deciding whether it is the most appropriate approach for any given task. A thorough analysis of the problem, and considering alternative approaches, is always prudent.

2.1 Procedural Approach

A procedural approach structures a program as a sequence of commands or procedures. Think of it like a detailed recipe: each step, precisely defined, leads to the next, ultimately culminating in the finished dish – your program's output. This methodical, step-by-step execution is its core strength, making it easy to understand, especially for beginners. The program's state changes directly as a result of these commands. Imagine a simple calculator program: each button press triggers a procedure to update the displayed value. This direct manipulation of data is a defining characteristic.

However, this linear flow can become unwieldy in complex projects. As the number of procedures grows, managing interactions between them becomes a significant challenge. Maintaining clarity and preventing unintended consequences requires meticulous organization and careful planning. Consider a large software application; managing the hundreds or even thousands of procedures necessary for its functionality would quickly overwhelm you. A linear structure, while fundamentally clear, struggles with complexity. Proper documentation, therefore, is critical.

This brings us to the importance of modularity in procedural programming. Breaking down large tasks into smaller, manageable procedures, each focusing on a specific subtask, enhances readability and maintainability. This is analogous to breaking a complex recipe into smaller, logical units—preparing the vegetables, mixing the sauce, and assembling the final dish. Well-defined interfaces between these modules greatly facilitate testing and debugging. Each procedure operates independently of others unless their interfaces explicitly state otherwise.

Functions, often used interchangeably with procedures, play a pivotal role in achieving modularity. Functions encapsulate a specific set of operations, taking inputs and returning outputs. This abstraction hides implementation details, allowing developers to focus on the what, not the how. This principle reduces complexity. Imagine using a pre-built function to sort an array; you don't need to understand the sorting algorithm to use it. You just provide the input and obtain the sorted result, improving code reusability.

Data structures are equally crucial. They organize and manage the data a program manipulates. Arrays, lists, and records provide different ways to store and access data, each suited to particular needs. Choosing the right data structure can significantly impact a program's efficiency and performance. A poorly chosen data structure can lead to unnecessary processing time and reduced efficiency, even with well-structured procedures. Careful consideration here will streamline your entire program's operation.

Procedural programming excels in situations needing clear, step-by-step execution, like controlling robotic arms or processing numerical data. Its simplicity is invaluable when learning to program, fostering fundamental programming concepts. However, its limitations in tackling complex systems become apparent as projects grow. While efficient for specific tasks, its inherent linear structure poses scalability challenges in complex software projects. The lack of inherent mechanisms for abstraction, such as those found in object-oriented programming, can lead to code that's difficult to maintain and modify. Understanding its strengths and limitations will prove invaluable as you choose the right paradigm for your projects. This structured approach, with its well-defined procedures and functions, offers a strong foundation, especially for beginners in programming. Mastering these concepts forms an essential step toward understanding more advanced paradigms. Its simplicity allows focus on basic programming structures and concepts, laying the groundwork for future exploration of more complex programming techniques and paradigms. Remember, the strength of procedural programming is not in its scalability, but in its clarity and ease of understanding, especially for beginners.

2.2 Structured Programming

Structured programming is a method of organizing code into logical blocks, enhancing readability and maintainability. It contrasts with earlier, less organized approaches where code flow could be difficult to follow, leading to errors and making modifications a nightmare. Think of it like building a house: instead of throwing bricks together haphazardly, structured programming provides a blueprint, ensuring a strong and stable structure.

This systematic approach utilizes three primary control structures: sequence, selection, and iteration. Sequence simply means executing statements one after another, in the order they appear. Selection, on the other hand, involves making decisions based on conditions, using constructs like if-else statements to choose different paths of execution depending on whether a condition is true or false. This allows for branching logic, making the program adaptable to varying inputs and situations. Imagine a program that determines eligibility for a loan – it uses selection to check income, credit score, and other factors before making a decision. Iteration, the third control structure, allows the repeated execution of a block of code. This is achieved through loops like for and while loops. Iterative structures are crucial for tasks involving repetitive actions, such as processing lists of data or performing calculations multiple times. For example, a program that calculates the average of a set of numbers uses a loop to iterate over each number, adding it to a sum before finally dividing by the total count.

The power of structured programming lies in its simplicity and clarity. By breaking down complex tasks into smaller, manageable modules, it reduces the complexity and improves overall code organization. This modularity makes debugging significantly easier, as errors can be isolated and corrected within specific modules without affecting the entire program. It also facilitates teamwork, as different programmers can work on separate modules concurrently, integrating their code seamlessly when completed. Furthermore, well-structured code is easier to understand and maintain, reducing the long-term cost of software development and maintenance.

Consider the alternative: unstructured programming, where the code flows arbitrarily. This often results in what's known as "spaghetti code"—a tangled mess of instructions with convoluted branching and jumps that are nearly impossible to understand or modify. Debugging becomes a Herculean task, with small changes potentially leading to unforeseen consequences elsewhere in the program. This not only increases development time but also makes it exceedingly difficult to ensure the software's correctness and reliability. Structured programming avoids this pitfall by enforcing a disciplined approach, enhancing program readability and reducing the risk of errors.

Beyond the basic control structures, employing functions and subroutines is also a cornerstone of structured programming. These are self-contained blocks of code that perform specific tasks. Functions promote code reusability by encapsulating common operations, which can be called multiple times from different parts of the program without rewriting the same code repeatedly. This modular approach also improves readability, as the program's logic is divided into smaller, more easily digestible units, making it easier to understand the overall functionality. Consider a program that needs to perform a complex mathematical calculation at several points: rather than repeating the calculation each time, it can be encapsulated in a function and called whenever needed.

Moreover, the use of appropriate data structures further enhances the effectiveness of structured programming. By organizing data efficiently using arrays, linked lists, stacks, or other suitable structures, you improve the program's performance and clarity. Selecting the right data structure for a given task significantly influences the program's overall efficiency. A poorly chosen data structure might lead to slow execution times, especially when dealing with large datasets. For example, searching for an element in an unsorted array is inefficient, while using a sorted array or a more sophisticated data structure, like a hash table or a binary search tree, allows for quicker searching.

In conclusion, structured programming is not merely a stylistic choice; it's a fundamental approach to software development that directly impacts the quality, maintainability, and efficiency of your programs. By adhering to its principles, you create code that is easier to understand, debug, and maintain, leading to more robust and reliable software. Mastering structured programming is crucial for any aspiring programmer, forming the foundation for more advanced programming paradigms and techniques. It is a cornerstone of sound programming practice, regardless of the programming language being used. Its benefits extend beyond the initial development phase; they continue to pay dividends throughout the software's lifecycle, ensuring its longevity and adaptability to future changes and requirements. This fundamental understanding is key to effectively leveraging the power of programming in solving real-world problems.

2.3 Control Structures

Sequential execution is the cornerstone of imperative programming. Imagine a recipe: you follow the steps one after another, precisely as written. Similarly, in imperative programs, instructions are executed linearly, unless explicitly altered by control structures. These structures govern the order of execution, allowing for decision-making and repetition, thereby enabling complex tasks to be broken down into manageable units. Without them, programs would be limited to simple, sequential operations, severely restricting their functionality. They are the vital components that make programs adaptive and robust.

Conditional statements, such as if, else if, and else, introduce decision-making power into a program. They evaluate a condition, and based on its truth value (true or false), execute a specific block of code. This ability to choose different execution paths based on runtime data is essential for creating flexible and dynamic applications, allowing your software to respond to various situations. For instance, consider a program that checks user input; conditional statements ensure that the program handles valid and invalid inputs differently, preventing unexpected crashes. The careful construction of these statements is paramount to preventing logical errors.

Looping constructs are another powerful tool for controlling program flow. They enable repetitive execution of a block of code as long as a condition remains true. The while loop continues until its condition becomes false, while the for loop iterates a specific number of times or over a range of values. This iterative capability is critical for tasks involving processing large datasets, such as calculating the sum of numbers in a list, iterating over a collection of user accounts, or performing calculations on each element of a matrix. These loops enhance efficiency significantly by avoiding code redundancy. They enable the concise processing of repeated tasks. Incorrectly implemented loops can, however, cause infinite loops, a common programming error resulting in the program getting stuck.

Nested control structures, where one control structure is placed inside another, allow for the creation of increasingly complex program logic. Imagine a program that processes a list of files. The outer loop iterates through the files, and the inner loop processes each line within each file. This nested structure effectively combines conditional execution and repetition, enabling sophisticated operations. Careful planning of nested structures is crucial to enhance readability and prevent unintended complications or errors. For example, deeply nested loops can create a significant performance bottleneck, something to carefully consider during development.

The switch statement, or its equivalent, offers a structured way to select one block of code from multiple options based on the value of an expression. This is often more efficient than a series of if-else if statements when dealing with many distinct possibilities. For instance, when evaluating a user's choice from a menu with many options, a switch statement allows for clearer and potentially faster code execution. This clarity enhances the maintainability of the code, making it easier for other programmers (and even your future self!) to understand and modify.
Exception handling provides a mechanism to gracefully manage runtime errors. Using try-catch blocks (or their equivalents), programs can anticipate and handle potential problems, such as file not found errors or division by zero errors. Instead of crashing upon encountering an error, the program can catch the exception, log the issue, and continue executing or take alternative actions. This robustness is critical for creating reliable and user-friendly applications. Well-designed exception handling makes for a more resilient and user-friendly program, improving the overall user experience.
Proper use of control structures is not just about syntax; it directly impacts the program's efficiency, readability, and maintainability. Inefficient use, such as unnecessary nested loops, can lead to slow execution, especially when dealing with large datasets. Poorly structured code using complex control flows without clear organization can become difficult to understand and maintain. Prioritizing clear, concise code, appropriately

employing control structures, and consistently adhering to good coding practices makes it more manageable. This is of paramount importance as it affects both initial development time and subsequent maintenance.
Understanding control structures is foundational to mastering imperative programming. They offer the power to control the flow of execution, allowing you to write programs that are flexible, efficient, and robust. Mastering them is paramount to becoming a proficient programmer. These structures are the essential tools you will use every day. Remember, practice is key. Experiment with different types of control structures, implement them in your own programs and, most importantly, always aim for clarity and efficiency in your code. The more you practice, the more intuitive and efficient your code will become. The ability to master these structures is the foundation for more advanced programming skills and constructs.

3 Object-Oriented Programming

Everything revolves around objects in object-oriented programming (OOP). Think of them as blueprints for creating instances, each with its own data and behaviors. A "class" defines this blueprint, detailing the characteristics (attributes) and actions (methods) an object will possess. For example, a "Car" class might have attributes like color, model, and speed, and methods such as accelerate, brake, and turn. Each individual car you create would be an "object" based on that "Car" class blueprint. This modularity fosters reusability and maintainability, simplifying complex projects.

Inheritance is a powerful OOP mechanism. It allows you to create new classes (child classes) based on existing ones (parent classes), inheriting their attributes and methods. This promotes code reuse; you don't have to redefine common functionalities for each new class. A "SportsCar" class could inherit from the "Car" class, adding attributes specific to sports cars like horsepower and handling, while still retaining the basic car functionalities. This structured approach improves efficiency, reducing redundant coding. Imagine building a library of pre-defined classes that you can reuse for different programs. That's a core benefit of inheritance. Polymorphism, meaning "many forms," allows objects of different classes to respond to the same method call in their own specific ways. For example, both "Car" and "SportsCar" objects might have a "drive()" method. However, the "drive()" method would behave differently for each—a sports car might accelerate faster. This adaptability is crucial for creating flexible and extensible software; you can easily add new classes without modifying existing code significantly. Consider a system handling various vehicle types – polymorphism enables seamless integration of new designs without extensive restructuring.

Encapsulation bundles data and methods that operate on that data within a class, hiding internal details and exposing only necessary information through a well-defined interface. Think of it as a protective shell around your class's data, preventing accidental modification or direct access. This protective layer enhances security, maintainability, and modularity, allowing modifications without impacting other parts of the code. This reduces unexpected behaviors and simplifies debugging by confining problems to specific areas within your class.

Abstraction focuses on showing only essential information and hiding unnecessary complexity. It simplifies interactions by presenting a high-level view of objects, hiding the underlying implementation details. For example, when you drive a car, you don't need to know the intricacies of the engine; you only need to know how to steer, accelerate, and brake. This concept is fundamental for creating user-friendly interfaces and manageable codebases. Complex systems become easier to comprehend and maintain through this simplification.

Design patterns are reusable solutions to common software design problems. They provide blueprints for structuring classes and objects, promoting code reusability, maintainability, and efficiency. Common patterns include Singleton (ensuring only one instance of a class exists), Factory (creating objects without specifying their concrete class), and Observer (defining a one-to-many dependency between objects). These patterns streamline the development process, guiding developers toward well-structured and robust designs, mitigating common programming pitfalls and improving team collaboration. Understanding and applying design patterns can dramatically increase your code quality.

Mastering OOP principles empowers you to write cleaner, more efficient, and maintainable code. By understanding classes, objects, inheritance, polymorphism, encapsulation, abstraction, and design patterns, you can build robust and scalable applications, which are increasingly crucial in the complex world of modern software development. It's about developing a structured and organized approach to programming. It's not simply a paradigm shift; it's a methodology that fundamentally transforms how you approach and solve software problems.

3.1 Classes and Objects

Imagine a blueprint for a house. This blueprint details every room, its dimensions, and the placement of doors and windows. In object-oriented programming (OOP), a class is like that blueprint. It defines the structure and behavior of objects, which are the actual houses built from that blueprint. Classes act as templates, providing a framework for creating objects with specific characteristics and actions. Think of it as a design pattern, one that's incredibly useful for building complex and reusable software. They are the fundamental building blocks of OOP, enabling programmers to create well-organized, flexible, and easily maintainable code. Without understanding classes, you are missing the cornerstone of object-oriented programming and wouldn't be able to construct robust and sophisticated applications. A class defines attributes, also known as data members or instance variables, which represent the state of an object. For example, in our house blueprint analogy, the attributes might include the number of rooms, the square footage, and the address. Classes also define methods, which are functions that specify the actions an object can perform. These actions can range from simple tasks to complex operations, depending on the design. Returning to our analogy, methods could be "open door," "turn on lights," or even "calculate property taxes." Methods allow the objects to interact with each other and with the outside world, making them active participants in a program's execution. These attributes and methods together define the complete structure and functionality of an object, ensuring well-structured code that's easy to manage.

Let's consider a practical example: a class representing a "Car." Its attributes could include "make," "model," "year," and "color." Its methods might be "start," "accelerate," "brake," and "turn." Each individual car object – your red 2023 Toyota Camry or your blue 2010 Honda Civic – is an instance of the "Car" class. Each of these instances will have its own specific values for the attributes (your car's specific make, model, and year, etc.), but they all share the same methods and basic structure. Therefore, creating multiple car objects is made efficient by having a single class definition.

Creating objects, or instances, from a class involves a process called instantiation. This is the act of bringing a class's blueprint to life, constructing a specific object based on that blueprint. During instantiation, memory is allocated to store the object's data (attributes), and the object is made ready for use. Instantiation is a fundamental step in the lifecycle of an object, taking a class's abstract definition and turning it into a tangible entity within a running program. The instantiation process itself can be as simple as calling the class's constructor, which is a specialized method responsible for creating the object and initializing its attributes.

Classes promote code reusability. Once a class is defined, you can create multiple objects from it, avoiding

redundant code. This is analogous to using a cookie cutter: you have one cutter (the class), but you can create many cookies (objects) from it. This reusability dramatically reduces development time and effort, fostering efficient and maintainable code structures. In larger projects, this attribute becomes particularly invaluable, reducing the chance of inconsistencies and facilitating easy modifications. The modularity and organization introduced by classes contribute to a much smoother development process.
Understanding classes and objects is crucial for mastering OOP. They provide a powerful mechanism for organizing code, managing complexity, and building robust and flexible software systems. They facilitate better code design and offer advantages like code reusability and easier maintenance, making them a crucial component of software development, especially in larger projects. It's more than just understanding the syntax; it's about grasping the underlying concepts and applying them effectively to build elegant and efficient programs. Without them, larger projects would quickly become unmanageable, and code maintainability would be a significant problem. So, take the time to fully grasp these fundamental concepts, and you'll find your software projects greatly benefit.

3.2 Inheritance & Polymorphism

Inheritance allows a class to inherit attributes and methods from another class, establishing an "is-a" relationship. This promotes code reusability and reduces redundancy. For example, a "Car" class might inherit from a more general "Vehicle" class, inheriting properties like number of wheels and a method for starting the engine. Subsequently, the "Car" class can add its own unique attributes, such as the number of doors or the type of transmission, demonstrating a clear hierarchical structure. This inheritance mechanism is fundamental to building complex, well-organized systems. We can visualize this relationship as a tree, with the base class forming the root and its subclasses branching out, each inheriting and extending the functionality of their parent. This approach significantly simplifies the development process by avoiding repetitive coding and fostering consistency. The advantages are substantial; think about the effort saved by not having to redefine common characteristics for every new vehicle type. Think of it this way: without inheritance, every single object would need to be defined from scratch, creating enormous complexity and increasing the likelihood of errors. Polymorphism, on the other hand, allows objects of different classes to respond to the same method call in their own specific way. This flexibility is crucial for creating adaptable and extensible software. Consider a "draw()" method. A "Circle" object might implement this method to draw a circle, while a "Square" object would implement it to draw a square. The calling code doesn't need to know the specific type of object; it just calls "draw()" and the correct method is executed based on the object's type. This dynamic behavior allows for seamless integration of various object types within a single system, offering a significant advantage in managing diverse components. For instance, a graphics library might use polymorphism extensively, enabling objects of different shapes to be drawn with a single, uniform interface. This reduces the amount of code needed to handle the diversity of shapes, ultimately enhancing the overall system's efficiency and maintainability. The elegance of this approach lies in its ability to handle complexity gracefully and abstract away the specifics of each object's implementation.

The combination of inheritance and polymorphism creates powerful and flexible object-oriented systems. Inheritance establishes a hierarchical structure, promoting code reuse and organization. Polymorphism then allows objects from different branches of this hierarchy to interact seamlessly through a common interface. This allows developers to build more complex systems with less code and increased maintainability. Imagine a scenario where you're designing a game with various characters. Using inheritance, you could create a base "Character" class with common attributes like health and attack power. Then, you could create subclasses for specific character types, like "Warrior," "Mage," and "Archer," each inheriting from the base class and adding their unique characteristics. Polymorphism would allow you to have a single "attack" method in each subclass, with each subclass defining its own unique implementation of how that attack works, maintaining a consistent interface while accommodating diverse behaviors. The overall result is a cleaner, more scalable, and efficient game architecture.

Consider, for instance, a banking application. We might have a base class "Account" with common attributes such as balance and account number. Then, we could create subclasses like "SavingsAccount" and "CheckingAccount," inheriting from the "Account" class but implementing different interest calculations or transaction fee structures. This organization is not only well-structured, but it also promotes maintainability; changes to the core "Account" class will automatically propagate to all subclasses. The ability to add new account types without altering existing code, due to polymorphism, illustrates the power of this combined approach. This approach is superior because it allows for simple extensibility without disrupting the overall design. Furthermore, it encourages code reuse, reducing the risk of errors introduced through duplicated code.

Let's take another practical example, focusing on graphical user interfaces (GUIs). A base "Button" class could define common features like size, position, and event handling. Subclasses like "TextButton" and "ImageButton" can then inherit these features and add their specific visual and functional attributes, such as displaying text or an image. The polymorphism comes in when the GUI system handles clicks on any button; it doesn't need to know the specific type of button, only that it's a button, triggering the appropriate action. This approach simplifies development considerably; the GUI framework only needs to deal with a generic button interface, regardless of the button's visual representation or behavior. The advantages of this design are substantial; it promotes modularity, allowing for independent development and evolution of different button types. Changes to the base class won't necessitate extensive modification of the subclasses, resulting in a more robust and maintainable system.

In conclusion, inheritance and polymorphism are powerful tools in object-oriented programming. Mastering these concepts significantly improves a programmer's ability to write efficient, reusable, and easily maintainable code. By understanding how these principles work together, you can create robust and scalable software solutions that can adapt to changing requirements more easily. The benefits extend beyond just writing efficient code to include easier debugging and expansion of the application's functionality in the future. This is a critical step in developing professional, industrial-strength software. Through the thoughtful application of inheritance and polymorphism, you gain a powerful framework for constructing modular and flexible systems. This understanding forms the cornerstone of creating truly effective and scalable software.

3.3 Encapsulation

Data hiding, a cornerstone of object-oriented programming, is what encapsulation truly embodies. It's not just about bundling data and methods together; it's about controlling access to that data, ensuring its integrity and preventing unintended modifications. Think of it like a well-guarded vault, protecting valuable assets from unauthorized access. The vault itself—the class—contains both the treasure (data) and the combination (methods). Only those with the correct combination (authorized methods within the class) can access and manipulate the treasure.

This controlled access is achieved through access modifiers, keywords that determine the visibility and accessibility of class members (data and methods). For instance, in languages like Java or C++, we have public, private, and protected members. Public members are freely accessible from anywhere, like an open door to a public park. Private members, conversely, are only accessible within the class itself, akin to that highly secured vault, protecting data from external interference. Protected members offer a middle ground, allowing access within the class and its subclasses, like restricted access to a members-only club. Careful consideration of access modifiers is crucial for building robust and secure applications.

Consider the example of a bank account class. The account balance is critical data and should be protected from arbitrary changes. Encapsulation ensures this by declaring the balance as private. Public methods like deposit() and withdraw() provide controlled access, validating transactions and preventing the balance from being manipulated directly. This approach ensures data consistency and protects the integrity of the account. Without encapsulation, direct access to the balance could lead to errors, security vulnerabilities, and data corruption. Imagine the chaos if anyone could directly alter the balance!

Beyond data protection, encapsulation enhances code maintainability and reusability. By hiding implementation details, you create a clean, well-defined interface that simplifies interaction with the class. Changes to the internal workings of the class don't necessarily require modification of code that uses it, as long as the public interface remains consistent. It's like upgrading the engine of a car—as long as the steering wheel and pedals remain in the same place, the driver's experience doesn't change. This decoupling of interface and implementation is a key advantage of encapsulation, simplifying code modifications and reducing the risk of unintended side effects.

Furthermore, encapsulation fosters modularity, a crucial aspect of large-scale software projects. Encapsulated classes act as independent, self-contained units, promoting better organization and easier collaboration among developers. Each class can be developed and tested in isolation, simplifying the overall development process. It is similar to assembling a complex machine using individual, well-defined components. Each component, or class, has its own function, and they work together to achieve a larger goal. This structure prevents widespread changes when modifications are needed to a specific class, unlike the monolithic approach of older code.

Proper encapsulation isn't just about applying access

modifiers; it's a design principle. It requires careful planning and thoughtful consideration of which members should be public and which should be kept private. Overusing public members can lead to tight coupling and reduced flexibility, negating the benefits of encapsulation. Underusing them can create unnecessary complexity. Finding the right balance is crucial for effective code design. A good rule of thumb is to expose only what's necessary for the external world and keep the internal workings hidden, improving security, simplifying maintenance, and promoting a better system structure. The goal is to create clean, robust, and maintainable software.

Effective encapsulation leads to more robust and reliable software, reducing the likelihood of errors and simplifying debugging. Because data is protected from external interference, the possibility of data corruption is minimized, reducing the time spent on fixing problems. Furthermore, the modularity promotes easier testing of individual components, leading to quicker identification and resolution of any issues. This enhances development speed and improves the overall quality of the resulting code.

3.4 Abstraction

Data hiding, the cornerstone of abstraction, isn't merely about concealing implementation details; it's about creating a clear boundary between what a piece of code does and how it does it. Imagine building a car: you interact with the steering wheel, the gas pedal, the brakes – you don't need to understand the intricate mechanics of the engine, transmission, or braking system to drive. Abstraction provides that same user-friendly interface in software. This simplifies interaction and reduces complexity, allowing developers to focus on higher-level logic without being bogged down in low-level details. The internal workings remain hidden, providing a level of insulation against changes in those underlying components. This is crucial for maintainability; modifying the engine shouldn't require rewriting the steering system.

Abstraction promotes modularity, a vital principle in large-scale software projects. By creating abstract interfaces, we decompose complex systems into smaller, more manageable modules. Each module has a well-defined purpose and interacts with other modules through clearly defined interfaces. This modularity simplifies testing, debugging, and modification. Changes within one module are less likely to have cascading effects on other parts of the system, improving the robustness and maintainability of the entire software. Think of it as constructing a building with prefabricated components—each component serves a specific function, and the overall structure remains unaffected by changes to individual parts, making upgrades and repairs easier and more efficient.

Consider the concept of an abstract data type (ADT). An ADT defines a set of operations on data, without specifying how the data is actually represented or the operations are implemented. For instance, a stack ADT might define operations like push and pop, but the underlying implementation (array-based or linked list-based) is hidden from the user. This allows different implementations to be swapped out without affecting the code that uses the stack, as long as the interface remains consistent. This is a powerful technique, promoting flexibility and code reusability. Using ADTs, different developers can work independently, creating modular software components that integrate effortlessly without having to worry about each other's specific implementation choices.

Furthermore, abstraction significantly enhances code readability and understanding. By hiding unnecessary complexity, it allows developers to focus on the essential aspects of the program's logic. Code that is easier to understand is easier to maintain, debug, and extend. This contributes to improved developer productivity and a reduction in errors. When code is cluttered with unnecessary implementation details, understanding its functionality becomes a daunting task. Abstraction helps to eliminate this clutter by providing a cleaner, more intuitive interface. Clean and simple code is the hallmark of any successful software project, and abstraction plays a vital role in achieving this goal.

Abstraction is not just a theoretical concept; it's a practical tool that directly affects the efficiency of the development process. By simplifying complex systems, it reduces the cognitive load on developers, allowing them to work more efficiently and effectively. This translates to faster development cycles, reduced development costs, and improved software quality. The time saved in understanding and modifying code can be substantial, especially in large projects where many developers collaborate. Moreover, the reduced error rate contributes to a more stable and reliable software product. This enhances the overall value of the software by reducing the costs associated with maintenance and bug fixes. A simpler, more understandable system is a more efficient one.

However, over-abstraction can be detrimental. While abstraction simplifies complexity, excessive abstraction can introduce its own kind of complexity, making the system harder to understand and maintain. Finding the right balance between abstraction and concrete implementation is crucial for successful software design. This requires careful consideration of the system's requirements and the trade-offs involved. Overly abstract systems can become difficult to implement and debug, leading to increased development time and costs. Therefore, it is essential to strive for a level of abstraction that simplifies the system without obscuring its underlying workings unnecessarily.

In conclusion, mastering abstraction is fundamental for writing efficient, maintainable, and scalable software. By carefully applying abstraction principles, we can create software systems that are easier to understand, modify, and debug. It's a crucial technique for managing complexity, promoting modularity, and enhancing the overall quality of our software projects. The ability to effectively utilize abstraction sets apart proficient programmers from novices, transforming software development from a chaotic process into a structured and efficient undertaking. Abstraction is not merely a feature; it's a cornerstone of effective software engineering.

3.5 Design Patterns

Design patterns are reusable solutions to common problems in software design. They're not specific code snippets, but rather templates that guide you in structuring your code. Think of them as blueprints for solving recurring design challenges, ensuring consistency and maintainability within your object-oriented projects. They help prevent reinventing the wheel every time a similar problem arises, leading to more efficient and reliable software development. Mastering these patterns significantly improves your coding skills and ultimately your ability to build robust applications.

Understanding design patterns requires grasping the core principles of object-oriented programming, specifically inheritance, polymorphism, and encapsulation. These three pillars form the foundation upon which many patterns are built. Without a solid understanding of these principles, the nuances and advantages of design patterns may remain elusive. It's crucial to familiarize yourself with these fundamental concepts before diving into the world of design patterns. Consider revisiting the earlier chapters of this book if you need a refresher.

One of the most prevalent patterns is the Singleton pattern. This pattern ensures that a class has only one instance and provides a global point of access to it. This is incredibly useful for managing resources that should only exist once, like a database connection or a logging service. The Singleton pattern promotes controlled access and prevents the creation of multiple instances that could lead to conflicts or resource depletion. Implementing a Singleton might involve using a private constructor and a static method to return the single instance. However, be mindful of potential challenges with testing and maintainability, which can arise if not implemented correctly.

The Factory pattern provides an elegant solution for creating objects without specifying their concrete classes. This proves extremely beneficial when dealing with many different types of objects that share a common interface. Instead of directly instantiating objects, you use a factory method that returns the correct object based on some criteria. Imagine a game where you need to create different types of enemies – each with unique characteristics. A Factory pattern allows you to create enemies without needing to know their precise types upfront, enhancing flexibility and extensibility. The Factory Method adds a layer of abstraction, hiding the instantiation process from the client code.

Another crucial pattern is the Observer pattern, dealing with the problem of one-to-many dependencies. This pattern establishes a one-to-many dependency between objects so that when one object changes state, all its dependents are automatically notified and updated. A classic example is a spreadsheet program, where updating a cell automatically recalculates dependent cells. The Observer pattern separates the subject (the object being observed) from its observers, making the code more modular and easier to maintain. This pattern greatly improves code organization and reduces complexity in scenarios involving dependent objects.

The Strategy pattern allows defining a family of algorithms, encapsulating each one, and making them interchangeable. This promotes flexibility by allowing the algorithm to be selected at runtime. Consider the scenario of a sorting algorithm; instead of hardcoding one particular algorithm, the Strategy pattern allows you to choose between various sorting algorithms based on the specific needs of the application. This adds significant adaptability to your program, enabling you to easily switch between different implementations without modifying the core structure of the application.

Finally, the Adapter pattern facilitates the collaboration of otherwise incompatible classes by converting the interface of a class into another interface that clients expect. Imagine needing to integrate a legacy system with a modern system; their interfaces are likely different. The Adapter pattern allows you to reconcile these discrepancies, making it possible to use the legacy system with the new system without extensive refactoring. It's a powerful tool for integrating components that were not initially designed to work together, promoting code reusability and minimizing code changes in existing systems. This pattern is particularly useful when dealing with third-party libraries or integrating legacy systems. Understanding and applying these fundamental design patterns will dramatically improve your programming efficiency and the overall quality of your software projects. Remember that design patterns are tools; choose the right one for the task at hand.

4 Functional Programming

Pure functions are the cornerstone of functional programming. They always produce the same output for a given input, and crucially, they have no side effects. This predictability makes debugging significantly easier; you can test a pure function in isolation, confident that its behavior will remain consistent regardless of the program's overall state. Understanding this fundamental principle is essential for grasping the power and elegance of this paradigm. This contrasts sharply with imperative programming, where functions often depend on and modify global variables, introducing complexity and making testing and reasoning about the code far more challenging.

Immutability, closely tied to pure functions, is another defining characteristic. Once a data structure is created, it cannot be changed. Instead of modifying existing data, functional programs create new data structures incorporating the changes. This approach might seem inefficient at first glance; however, it greatly simplifies concurrency. Because data is immutable, multiple parts of a program can access and process it simultaneously without the risk of race conditions or data corruption—a significant advantage over imperative methods where shared mutable state often requires intricate synchronization mechanisms. Consider the simple act of appending to a list; in an imperative language, this modifies the original list directly. In a functional setting, you'd create a completely new list containing the original elements plus the new one. This subtle shift results in cleaner, less error-prone code.

Higher-order functions are functions that take other functions as arguments or return them as results. This capability allows for abstraction and code reuse in powerful ways. Imagine a function that sorts a list; in a functional language, you could write this function to accept a comparison function as an argument. This allows you to easily sort the list using different criteria without modifying the sorting function itself. This flexibility is a core strength of functional programming, making the code more adaptable and reusable. It promotes a style of programming where you compose functions together, building complex behavior from simpler, more manageable parts. This contrasts significantly with the more explicit, step-by-step approach of imperative programming.

Recursion, the ability of a function to call itself, is a powerful tool in functional programming. Many algorithms that would be expressed iteratively in imperative languages are often expressed recursively in functional ones. For example, traversing a tree or computing factorials are elegantly handled using recursion. While recursion can sometimes lead to stack overflow issues if not managed carefully, its elegance and conciseness are often preferred by functional programmers, particularly when dealing with inherently recursive data structures. Understanding the base case and recursive step is crucial when designing recursive functions. Remember that each recursive call effectively breaks down the problem into smaller, self-similar subproblems until the base case is reached, at which point the results are combined to produce the final output.

Lazy evaluation is a technique where expressions are not evaluated until their values are needed. This can dramatically improve performance in some cases, especially when dealing with potentially expensive computations. If a value is never used, it's never computed, leading to significant efficiency gains. It allows for the creation of infinite data structures – such as infinite lists – that wouldn't be feasible with eager evaluation. While not always applicable, lazy evaluation can contribute greatly to creating efficient and responsive programs, particularly those involving complex data processing pipelines where the need for intermediate results might not always be apparent. Careful design and understanding of potential performance implications are still necessary to utilize lazy evaluation effectively. This feature, often found in functional languages, underscores how functional programming focuses on expressing what to compute rather than how to com-

pute it.
Functional programming is not without its challenges. The emphasis on immutability can sometimes lead to more memory consumption than imperative approaches, as new data structures are constantly created. The learning curve can also be steep for programmers accustomed to the mutable state and side effects of imperative programming. However, the benefits of increased code readability, maintainability, and testability often outweigh these challenges, especially in large, complex software projects where the advantages of concurrency and reduced complexity become paramount. The focus on pure functions and immutability promotes a style of programming that results in cleaner, easier-to-understand code, simplifying collaborative development and long-term maintenance. Ultimately, the choice between paradigms depends on the specific project's requirements and the strengths of the team. But understanding functional programming is essential for any serious software developer, providing a powerful alternative perspective on code design.

4.1 Pure Functions

The essence of functional programming lies in the concept of pure functions. These are functions that, given the same input, always produce the same output, and crucially, they have no side effects. This predictability is a cornerstone of functional programming's power. It eliminates the unexpected behavior often associated with functions that modify global variables or interact with external systems, leading to easier debugging and a more robust application. Think of a mathematical function: the square root of 9 is always 3, regardless of what calculations have occurred previously. Pure functions operate in a similar, predictable fashion.

Consider a simple example: a function that adds two numbers. If this function is pure, it solely relies on the input values provided and doesn't alter any external state. This contrasts with a function that might update a global variable or write to a file; these actions constitute side effects. Eliminating side effects drastically simplifies the process of testing and reasoning about your code. You can test a pure function by simply inputting various values and checking the result; you don't need to consider external dependencies or their current states.

Pure functions promote modularity and code reusability. Because they rely solely on their inputs, you can easily incorporate them into other parts of your program without worrying about unintended consequences. This reduces the risk of introducing bugs, and also makes the code easier to maintain and modify. If a section of code is dependent on a pure function, you can refactor or replace the function without affecting other parts of your program as long as the input/output behavior remains consistent. This modularity aids in larger project development, as it allows developers to work independently and seamlessly merge their code.

The benefits of utilizing pure functions extend beyond

simple debugging and code maintenance. They are particularly advantageous for concurrent programming. Because pure functions don't share state or have side effects, multiple threads or processes can execute them simultaneously without risking data corruption or race conditions. This characteristic significantly improves performance and enables the efficient utilization of multi-core processors. The absence of shared mutable state simplifies concurrency control mechanisms, reducing the complexity of handling thread synchronization.

However, relying exclusively on pure functions isn't always practical. Some operations inherently involve side effects; input/output operations are a prime example. Reading data from a file or updating a database are essential yet non-pure operations. However, a good practice is to confine side effects to minimal, clearly defined areas of your code, treating them as interfaces to the external world. By keeping your core logic based on pure functions, you can maximize the benefits of this approach while still handling necessary interactions with the external environment. Strive for the optimal balance; maximize pure functions wherever possible, and isolate side effects to strategic points for better control.

The transition to using pure functions might seem challenging at first, especially for programmers accustomed to imperative programming styles. It requires a shift in mindset, prioritizing immutability and avoiding the modification of global state. You'll find yourself increasingly breaking down complex operations into smaller, more manageable pure functions. Consider this an opportunity to elevate the clarity and precision of your code. This iterative approach allows for a gradual integration of functional programming techniques. Begin with small sections of your existing code. Refactor functions, breaking them into smaller, more focused ones and isolating side effects as clearly defined components. Remember, the goal is not to eliminate all side effects entirely. The focus is on minimizing and carefully managing them, leveraging the significant advantages of pure functions for the bulk of your program's logic. This pragmatic approach balances the theoretical elegance of pure functions with the practical necessities of software development. By strategically integrating pure functions, you improve code readability, maintainability, testability, and concurrency. Embrace the power of pure functions to transform your programming skills and produce more efficient and robust software. The initial investment in learning and adjusting your coding style pays off exponentially in the long run through cleaner code and increased productivity. Don't be afraid to experiment and gradually incorporate these principles into your projects. You'll soon appreciate the benefits of embracing this fundamental aspect of functional programming.

4.2 Immutability

Immutability is a cornerstone of functional programming, a paradigm that prioritizes the creation of functions that always produce the same output for the same input, without side effects. This principle stands in stark contrast to imperative programming where data is frequently modified in place. Understanding immutability is crucial for writing cleaner, more predictable, and easier-to-debug code.

In essence, an immutable object cannot be changed after its creation. Any operation that appears to modify an immutable object actually creates a new object with the desired changes. Consider a simple example: if you have an immutable list and want to add an element, the original list remains untouched; instead, a new list containing the added element is created. This might seem inefficient at first glance, but the benefits outweigh the potential performance overhead, especially in complex systems.

The advantages of embracing immutability are multifaceted. First, it significantly simplifies debugging. Since an immutable object's state never changes, you can confidently reason about its value at any point in your program's execution. You eliminate the risk of unexpected modifications causing subtle bugs that are notoriously difficult to track down. The absence of side effects, a direct consequence of immutability, further contributes to this predictability.

Testing also becomes considerably easier with immutability. Because functions operating on immutable data always produce the same output for the same input, tests become much simpler and more reliable. You don't need to worry about the state of the data changing between test runs, which dramatically reduces the complexity of testing and improves the overall quality of the code.

Furthermore, immutability dramatically enhances concurrency. In multi-threaded applications, multiple threads can access and operate on immutable data concurrently without the need for complex synchronization mechanisms. This is because there's no risk of data races or other concurrency-related issues that often plague mutable data. This streamlined approach significantly simplifies concurrent programming and leads to more efficient and robust code, especially in high-performance applications.

However, implementing immutability isn't without its challenges. A naive approach can lead to significant performance overhead, especially if you're frequently creating new objects. This is where techniques like structural sharing come into play. Structural sharing optimizes memory usage by sharing common parts of immutable data structures, thereby reducing the memory footprint and improving performance. Smart compiler optimizations also play a crucial role in mitigating performance penalties associated with immutability.

Another potential challenge is the initial learning curve. Programmers accustomed to imperative programming styles might find the shift to immutability initially difficult. It requires a change in mindset, focusing on transforming data rather than modifying it in place. This paradigm shift necessitates a deeper understanding of functional programming concepts such as pure

functions, higher-order functions, and function composition.
Despite these challenges, the benefits of immutability are compelling. It leads to more robust, maintainable, and scalable software. By reducing bugs, simplifying testing, and enabling easier concurrency, immutability is a powerful tool in a programmer's arsenal. It's not merely a stylistic choice; it's a fundamental design principle that can significantly improve the quality and efficiency of your programs. As you gain experience with functional programming, you'll discover that immutability is not just a technique, but a mindset that fosters better programming practices. Embracing immutability will significantly improve your code's quality and maintainability, leading to more efficient and reliable software. The initial investment in learning and adapting to this paradigm will pay significant dividends in the long run. Therefore, it's a skill worth cultivating for any serious programmer.

5 Logic Programming

Logic programming, unlike imperative or functional approaches, doesn't focus on how to solve a problem step-by-step. Instead, it concentrates on what the problem is, expressing it as a set of logical statements. These statements, composed of predicates and clauses, describe relationships between facts and rules. Imagine building with LEGOs: imperative programming would dictate each hand movement, functional programming would define functions to build specific parts, but logic programming would describe the final structure, letting the system figure out the construction steps. This declarative nature is a core strength, making programs easier to understand and reason about, particularly for problems easily represented logically, such as expert systems or theorem proving.

The foundation of logic programming is the predicate. A predicate is a statement that can be either true or false, often representing a relationship between objects. For example, likes(John, Mary) would represent the fact that John likes Mary. Predicates are combined with logical connectives like AND (&), OR (|), and NOT (\)

to form more complex expressions. These expressions are then organized into clauses, which are essentially implications – if certain conditions are met, then a conclusion follows.

Unification is the crucial process that allows a logic program to deduce new facts from existing ones. It works by comparing two predicates and attempting to find a way to make them identical by substituting variables with terms. This substitution process is called instantiation. Think of it like solving a puzzle; unification finds the pieces that fit together to create a complete picture. For instance, given the facts likes(John, X) and likes(X, Mary), unification finds the substitution X = Mary, establishing the new fact likes(John, Mary). This seemingly simple mechanism is the engine of logical deduction in logic programming.

Resolution is the process of combining clauses to derive new clauses, eventually aiming for a conclusion. It's a systematic approach to finding contradictions, which in logic programming means finding a proof. If you have a clause saying "If A is true, then B is true," and another saying "A is true," resolution infers "B is true." This deduction process continues until either a contradiction is found (proving the initial statement false) or a conclusion is reached, proving the initial statement true. In essence, resolution systematically explores the logical consequences of the facts and rules provided, searching for a proof or disproof of a given query. Understanding resolution is fundamental to grasping the inferential power of logic programming.

Prolog, the most well-known logic programming language, exemplifies these principles. In Prolog, facts are declared using predicates, and rules are defined using implications. A programmer queries the system by posing a question in the form of a predicate, and Prolog uses unification and resolution to deduce an answer, or determine whether the query is satisfiable. For example, if you have facts about family relationships, you can ask Prolog questions like "Who is John's father?" and it will use the established rules and facts to find the answer. This ability to deduce answers directly from logical statements makes Prolog well-suited for knowledge representation and reasoning tasks.

The power of logic programming lies in its declarative style. Instead of explicitly detailing the steps to solve a problem, you define the problem's logical structure. The system then uses its inference engine – based on unification and resolution – to automatically find solutions. This approach can greatly simplify program design and make programs more understandable and easier to maintain, especially for tasks with a strong logical component. However, it's also crucial to understand that logic programming isn't a silver bullet. Its effectiveness depends heavily on the ability to represent problems in a declarative, logical fashion. Not all problems are readily amenable to this paradigm.

Developing proficiency in logic programming requires practice in representing problems logically, designing effective predicates and clauses, and understanding the intricacies of unification and resolution. While it might initially seem different from imperative or functional programming, mastering this paradigm opens up new possibilities for tackling complex problems, especially those involving knowledge representation, reasoning, and automated deduction. It's a distinct perspective that expands the programmer's toolkit and opens doors to a powerful and expressive style of programming. The declarative nature, while sometimes requiring a shift in thinking, results in elegant solutions for appropriate problems, making it a valuable addition to any programmer's repertoire. Learning its nuances will undoubtedly enhance problem-solving abilities across various programming domains.

5.1 Predicates and Clauses

Logic programming offers a unique approach to problem-solving, diverging significantly from the imperative and functional paradigms. Instead of specifying how to solve a problem step-by-step, as in imperative programming, or defining functions transforming inputs into outputs, as in functional programming, logic programming focuses on what the solution should be. This is achieved through the use of logical statements, forming the very foundation of this paradigm. These statements, in essence, declare facts and rules about a domain, allowing the system to deduce new facts and ultimately find solutions.

The core building blocks of logic programs are predicates and clauses. Predicates are essentially relationships between objects, represented as logical assertions. For instance, likes(John, Mary) is a predicate asserting that John likes Mary. This is a simple, atomic fact. We can also have more complex predicates, involving multiple arguments and representing intricate relationships. Consider parent_of(X, Y), signifying X is a parent of Y. The variables X and Y allow for generality; this single predicate can represent many parent-child relationships. Understanding predicates is crucial because they form the backbone of knowledge representation within logic programming. The elegance lies in their declarative nature; you describe the relationships, and the system figures out the connections.

Clauses, on the other hand, express rules and implications. A clause is built from predicates, using logical connectives such as implication (=>) and conjunction (and). A simple example might be ancestor_of(X, Z) :\item parent_of(X, Y), ancestor_of(Y, Z). This translates to "X is an ancestor of Z if X is a parent of Y and Y is an ancestor of Z." This recursive definition elegantly captures the transitive nature of ancestry. Notice how this rule doesn't specify how to find ancestors; instead, it describes the conditions under which the ancestor relationship holds. This declarative style reduces complexity and makes programs easier to understand and reason about.

The power of logic programming lies in the system's ability to automatically infer facts from given rules and facts. This process, known as inference or resolution, involves matching patterns, applying rules, and generating new facts. Imagine we know parent_of(John, Mary) and parent_of(Mary, Susan). Using the ancestor_of rule defined earlier, the system can automatically deduce ancestor_of(John, Susan). This automatic deduction capabilities simplifies complex reasoning tasks. Consider a more intricate scenario, involving family relationships and properties. Let's define predicates like owns(Person, Object), indicating ownership, and expensive(Object), classifying objects as expensive. We could create clauses like rich(Person):-owns(Person, Object), expensive(Object). This states that a person is rich if they own an expensive object. This example demonstrates the ability to combine predicates and express more sophisticated relationships. You can define numerous predicates representing properties, actions and relationships, crafting a rich model of your specific domain.

The true beauty of this system emerges from its ability to handle uncertainty. Instead of strict true/false statements, logic programming allows for the use of negations and disjunctions, reflecting the complexity of real-world situations. You can express situations where something might be true or false, or might be true under certain conditions. This opens up possibilities for modeling scenarios with incomplete or uncertain information, extending the usefulness beyond simple deductive reasoning.

Furthermore, the use of variables within predicates and clauses enables powerful generalization. Instead of writing separate rules for each specific instance, we can use variables to represent generic cases. This feature allows for a concise representation of knowledge, leading to more efficient and scalable logic programs. The flexibility in defining relationships and applying rules results in a powerful tool for modeling diverse scenarios.

To truly master logic programming, hands-on experience is essential. Start with simple examples, gradually building complexity and experimenting with different predicates and clauses. Begin with clearly defined predicates, focusing on the relationships they represent, then move to constructing rules that logically connect these predicates. You can create simple programs to model family trees, knowledge bases, or even simple games. As you gain confidence, you can explore more advanced features such as negation and disjunction, expanding the expressiveness of your logic programs. Remember the key: clarity in defining predicates and rules is paramount for creating efficient and understandable logic programs. The elegance of logic programming shines in its ability to express complex problems with relative conciseness and simplicity, making it a compelling tool for specialized applications. Practice building your own knowledge representation using predicates and clauses to reinforce your understanding of this declarative style of programming.

5.2 Unification

Unification is a cornerstone of logic programming, a process that determines whether two logical expressions can be made identical by substituting variables with terms. Imagine it as a sophisticated puzzle: you have two expressions, potentially containing variables, and your goal is to find a set of replacements—the substitutions—that make them match exactly. This seemingly simple act underpins the entire reasoning mechanism of logic programs. Success means the expressions are unifiable; failure indicates they are inherently different and cannot be reconciled. This seemingly simple process opens the door to powerful deductive reasoning capabilities.

The mechanics involve a systematic comparison, working from the outermost structures inwards. If the expressions differ in their basic structure (e.g., a function call versus a variable), unification fails immediately. However, if they match, or if one is a variable, the process continues. Variable substitution is key. When a variable is encountered, it's assigned the value of its corresponding component in the other expression. This assignment is crucial; it is not a simple equality check; rather, it's a binding that dictates future behavior.

Consider two simple expressions: f(x, a) and f(b, y). To unify these, we need to find substitutions for x and y. The process might start by comparing the function symbols (f). These match, so we move to the arguments. x and b are compared. x being a variable, we can substitute x with b. Similarly, a and y are compared. We substitute y with a. The result is f(b, a), which is identical to both original expressions after the substitutions. The unification is successful, with the substitution set {x/b, y/a}.

But unification isn't always straightforward. Consider f(x, x) and f(a, b). Attempting to unify these, we'd initially substitute x with a. However, the next comparison reveals a conflict: the second x (already substituted with a) is compared with b, and a does not equal b. This results in a unification failure; these two expressions are fundamentally incompatible. It's important to note that the order in which variables are substituted can influence the outcome; choosing a different substitution order might make the process seem simpler.

Furthermore, circular substitutions, where the substitution involves the variable itself, must be handled carefully. Imagine trying to unify x with f(x). This would require substituting x with f(x), leading to an infinite regress: x becomes f(f(x)), then f(f(f(x))), and so on. Logic programming languages detect and prevent these circular substitutions, preventing infinite loops and ensuring the unification process remains computationally tractable. The avoidance of these loops is crucial for preventing a program from becoming unusable, as an infinite loop will stop any algorithm or process from ever ending.

The elegance of unification lies in its simplicity and power. It provides a clean, logical mechanism for matching patterns and deriving consequences. It's not merely a comparison process but a sophisticated rule-based transformation that underpins the inferencing power of Prolog and other logic programming languages. The unification algorithm forms the heart of the inference engine; it meticulously examines the expressions, making substitutions as needed, and promptly flags any inconsistencies. Mastering unification requires understanding its core principles, managing variable substitutions accurately, and acknowledging the limitations posed by conflicts and circular dependencies.

This process is fundamental to how logic programs solve problems. Given a set of facts (database of knowledge) and rules (logic relations between facts), the system applies unification to find solutions. A query, itself a logical expression, is compared against the facts and rules, applying unification to determine if there are valid substitutions that confirm the query's truth. This process of pattern matching and substitution forms the basis of logical inference. If a query matches a fact or a conclusion derived from the facts and rules through the repeated use of this system, a solution is found.

In summary, unification is not merely an algorithm; it's a fundamental principle that enables the power of logic programming. Its capability to systematically match and substitute elements within logical expressions provides a robust mechanism for logical inference and problem-solving. Understanding its mechanics—substitutions, conflict detection, and the avoidance of circularity—is crucial for anyone seeking to build or comprehend logic programs. It's a complex system built on a deceptively simple mechanism. The unification algorithm's effectiveness and speed are key to the runtime performance of logic programs, as complex programs can potentially undergo thousands of unification steps in order to reach a conclusion.

5.3 Resolution

Resolution is a powerful inference rule in logic programming, allowing us to deduce new facts from existing ones. It's the engine that drives the process of finding solutions within a logic program. Imagine you have a set of facts and rules, like pieces of a puzzle. Resolution helps you fit those pieces together to find a complete picture, a solution to your problem. Understanding resolution is key to grasping the power and elegance of logic programming.

Unlike imperative programming, which dictates precise steps, logic programming focuses on declaring what's true and letting the system deduce consequences. This declarative approach shifts the burden of how to solve a problem from the programmer to the inference engine. The programmer's task is to define the relationships between facts, leaving the engine to find the connections that lead to a solution. This is a fundamental shift in thinking, making it crucial for beginner programmers to absorb this concept thoroughly.

Consider a simple example: We have facts stating "Socrates is a man" and "All men are mortal." Resolution allows us to deduce the fact "Socrates is mortal."

The process involves unifying clauses (statements) to eliminate a shared variable, in this case "man." This act of unification, a core element of resolution, is a clever mechanism that enables powerful inferences. Through skillful unification, the system can find paths to the solution, not by brute force or complex algorithms but through logical deduction.

Let's delve deeper into the mechanics. Resolution operates on clauses written in conjunctive normal form (CNF), a standardized representation of logical statements. Each clause is a disjunction of literals (atomic propositions or their negations). The resolution rule states that if we have two clauses, one containing a literal and the other containing the negation of that literal, we can resolve them by combining the remaining literals into a new clause. This new clause is logically implied by the original two. This step-by-step approach allows for logical deduction and systematic problem-solving.

The power of resolution lies in its ability to systematically explore all possible inferences. It's a methodical process that ensures no potential solution is overlooked. This systematic exploration is crucial for finding solutions in complex problems. Imagine a large network of interconnected facts and rules. Resolution provides a structured way to traverse this network, systematically identifying paths that lead to a solution. This contrasts sharply with the more haphazard approach often found in imperative programming.

Implementing resolution requires a specific algorithm, often involving a search for complementary literals within a set of clauses. The search might employ techniques like backtracking or other sophisticated search strategies to ensure all possibilities are explored. Therefore, careful selection of algorithms and implementation techniques is crucial to maximize efficiency in large and complex knowledge bases.

A significant challenge beginners often face is understanding the distinction between resolution and other inference rules. While similar, resolution focuses on finding contradictions, whereas other rules might focus on direct implication. Mastering this distinction is critical for effectively applying resolution to problem-solving. This understanding enables programmers to create efficient logic programs. Mastering logic programming concepts can open up a wealth of opportunities in fields like artificial intelligence and database management.

Furthermore, the application of resolution is not limited to theoretical exercises. It forms the basis for many practical applications. Prolog, a popular logic programming language, directly utilizes resolution for its inference engine. Understanding how resolution works unlocks the power of Prolog and its potential for creating expressive and elegant programs. In essence, Prolog is a testament to the profound practical implications of resolution in real-world software development.

Finally, grasping resolution allows you to better appreciate the elegance and efficiency of logic programming. Instead of explicitly coding step-by-step instructions, you declare facts and rules, and the system deduces solutions. This declarative nature leads to cleaner, more maintainable, and often more efficient code, especially for problems well-suited to logical representation. Therefore, thoroughly understanding resolution is not just a theoretical exercise; it's essential for effective logic programming. The mastery of this crucial component significantly elevates one's programming skills and overall problem-solving ability.

6 Concurrent Programming

Concurrent programming tackles the challenge of performing multiple tasks seemingly simultaneously. This is crucial for maximizing performance, especially in applications demanding speed and efficiency, like video editing software or complex simulations. We achieve concurrency through threads and processes, distinct mechanisms with subtle but important differences. Threads, lightweight units of execution, share the same memory space within a process, making communication efficient but requiring careful synchronization to avoid conflicts. Processes, on the other hand, operate in their own isolated memory areas, offering improved security and stability at the cost of potentially slower inter-process communication.

Understanding the nuances of these execution models is vital. Threads are ideal when tasks share data and frequent communication is necessary; their shared memory enables quick data exchange. However, this shared access mandates meticulous synchronization to prevent race conditions—where multiple threads modify shared data simultaneously, leading to unpredictable results. Processes, while more resource-intensive due to their separate memory spaces, provide a safer environment, especially in handling error conditions, since a crash in one process doesn't necessarily affect others. Choosing between them hinges on your application's specific needs: performance versus robustness.

Synchronization mechanisms are fundamental in concurrent programming. They regulate access to shared resources, preventing race conditions and ensuring data integrity. Common methods include mutexes (mutual exclusion locks), semaphores, and monitors. Mutexes provide exclusive access to a resource; only one thread can hold the mutex at any given time. Semaphores, more versatile, control access to a resource based on a counter; threads can increment or decrement this counter, allowing a limited number of concurrent accesses. Monitors, higher-level abstractions, encapsulate shared data and the methods for accessing it, simplifying synchronization logic. Selecting the right mechanism depends on the complexity of your concurrency needs. A simple mutex might suffice for basic shared resource protection, while semaphores provide more flexible control for managing access to limited resources, and monitors offer a cleaner, more structured approach for complex scenarios. Careful consideration of these synchronization mechanisms is essential for building robust and reliable concurrent systems.

Deadlocks represent a particularly insidious problem

in concurrent programming. A deadlock occurs when two or more threads are blocked indefinitely, waiting for each other to release resources that are needed for their own progress. Imagine two threads, each holding a resource the other needs. Neither can proceed, leading to a complete standstill. Preventing deadlocks requires careful design and analysis of resource allocation. Techniques such as resource ordering (acquiring resources in a predetermined order to avoid circular dependencies), deadlock detection (monitoring system states to identify potential deadlocks), and timeout mechanisms (allowing threads to give up after a certain wait time) are crucial tools in mitigating this risk. Robust concurrent programs anticipate and handle potential deadlocks gracefully, rather than allowing them to disrupt functionality entirely. Proactive deadlock prevention, through careful resource management and algorithmic design, is far preferable to reactive solutions, which may still disrupt the flow of a program.

Beyond the core concepts of threads, processes, and synchronization, effective concurrent programming involves deeper considerations of program design. A well-structured concurrent program often employs strategies like thread pools to manage thread creation and reuse efficiently. Instead of creating and destroying threads for each task, a thread pool maintains a set of pre-created threads ready to handle incoming requests, reducing overhead. This optimization significantly improves performance in applications processing many concurrent requests, such as web servers. Efficient thread management also includes considering factors like thread priorities and scheduling policies. Strategic assignment of priorities allows higher-priority threads to receive more processing time, ensuring responsive behavior. Furthermore, understanding the operating system's scheduling algorithm can help predict and optimize the execution patterns of concurrent threads, improving overall system efficiency.

Finally, testing and debugging concurrent programs present unique challenges. The non-deterministic nature of concurrency makes reproducing bugs difficult; the same code can execute differently depending on timing and resource availability. Systematic testing methodologies, including stress testing (simulating high loads to identify concurrency bottlenecks) and specialized debugging tools (allowing step-by-step analysis of thread execution), are essential. Thorough testing ensures robustness and reliability in concurrent applications. The unpredictable behavior inherent in concurrency highlights the need for rigorous testing and debugging strategies far beyond what's typically sufficient for sequential programs. Remember: diligent testing and well-defined strategies for handling errors, such as exception handling and graceful degradation, are crucial in creating dependable concurrent systems.

6.1 Threads and Processes

Understanding the fundamental differences between threads and processes is crucial for any programmer venturing into concurrent programming. Processes are independent, self-contained units of execution, each possessing its own memory space, system resources, and execution context. Imagine them as entirely separate apartments in a large building; they're distinct and don't directly share resources. Threads, on the other hand, are lighter-weight entities that share the same memory space and system resources as other threads within the same process. Think of them as multiple residents within a single apartment; they share the common space but have their own individual tasks.

This inherent difference significantly impacts resource management. Processes, due to their isolation, enjoy greater security and stability. If one process crashes, others remain unaffected. The overhead of creating and managing a process, however, is higher. Threads, conversely, boast reduced creation and management overhead; their shared memory fosters faster inter-thread communication. This efficiency comes with a trade-off: shared memory necessitates careful synchronization to prevent data corruption or race conditions—a potential pitfall that requires rigorous programming practices.

Consider a scenario involving a web server. Each incoming request could be handled by a separate process. This approach ensures isolation; a malfunctioning request won't bring down the entire server. The cost is the higher overhead for managing numerous processes. Alternatively, the server could utilize threads, handling multiple requests concurrently within a single process. This would be far more efficient, however, meticulous synchronization mechanisms must be implemented to protect shared data, like session information or database connections. Improper synchronization can lead to unpredictable results, data inconsistency and potentially even system crashes.

Choosing between threads and processes depends on your specific needs. For applications where isolation and stability are paramount, processes are preferable, even at the cost of efficiency. When performance and rapid inter-thread communication are critical, threads offer a compelling advantage. However, careful consideration of synchronization is mandatory to prevent concurrency-related bugs. The level of complexity increases considerably when handling several threads, demanding sophisticated programming techniques to handle potential race conditions, deadlock situations and other forms of inter-thread communication issues.

Let's delve into a practical example illustrating the benefits of threads. Imagine you need to process a large dataset, performing the same operation on numerous subsets of data. Running these operations sequentially would be slow. Using multiple threads, each handling a subset, significantly speeds up the overall processing time. The threads share access to the common dataset, enhancing efficiency, yet this shared access requires synchronized access methods to prevent conflicts. Without appropriate synchronization, multiple threads could try to modify the same data simultaneously resulting in data corruption. Appropriate synchronization primitives, such as mutexes, semaphores, or monitors, are essential tools in a multi-threaded environment. These synchronization techniques regulate access to shared resources, preventing race conditions and ensuring data integrity.

The selection between threads and processes, therefore, isn't simply a matter of picking the faster option. It necessitates a deep understanding of your application's requirements and careful consideration of resource management and potential concurrency issues. A poorly designed multi-threaded application can be significantly less stable and more difficult to debug than a well-designed single-threaded or multi-process application. Effective synchronization mechanisms become crucial as the complexity of concurrent programs increases, requiring a more sophisticated approach to managing access to shared resources. For instance, deadlocks, a scenario where two or more threads are blocked indefinitely waiting for each other, represent a significant challenge in concurrent programming and require careful design and implementation of synchronization protocols to avoid.

Moreover, the choice can be further complicated by the operating system's underlying mechanisms. Some systems excel at managing numerous lightweight threads, while others are better suited for managing fewer, heavier processes. Thorough understanding of the target platform's capabilities is, thus, a necessary step in the process of deciding whether to opt for a multi-threaded or a multi-process architecture for your application. Finally, remember that the choice between threads and processes is often not an either/or decision; hybrid approaches, where both threads and processes are used strategically within a system, frequently prove optimal for complex applications requiring high performance and improved robustness. This demands advanced expertise in concurrency management and resource allocation, but its effectiveness can lead to highly efficient and resilient software solutions. The optimal solution often involves a blend of strategies tailored to the specifics of the program and the underlying system's capabilities. Choosing a multi-process architecture is sometimes preferred, even though it leads to a decrease in performance, because it avoids the complexity associated with dealing with concurrent data access, and because it simplifies debugging.

6.2 Synchronization

Threads, the fundamental building blocks of concurrent programming, offer the potential for significant performance gains. However, uncontrolled access to shared resources by multiple threads creates a minefield of potential problems. Imagine several cooks trying to simultaneously prepare a single dish; chaos ensues unless carefully orchestrated. This is precisely the challenge synchronization addresses. It's about establishing rules of engagement, ensuring that only one thread accesses a critical resource at any given time. Failing to do so can lead to data corruption, unpredictable behavior, and ultimately, program failure. The complexities involved can be subtle, demanding a deep understanding of the underlying mechanisms.

Synchronization mechanisms provide the necessary control. Mutexes, or mutual exclusion locks, are a common approach. Think of a mutex as a key to a shared resource; only the thread holding the key can access it. When a thread needs the resource, it tries to acquire the mutex; if the resource is free, it succeeds and gains exclusive access. Once finished, it releases the mutex, allowing other threads to contend for access. This ensures controlled access, preventing race conditions where the outcome depends on the unpredictable order of thread execution. The key is careful management, acquiring and releasing the mutex at precisely the right moments within your code.

Semaphores offer a more generalized approach. Unlike mutexes which manage binary access (locked or unlocked), semaphores allow for a variable number of simultaneous accesses. Imagine a parking lot with a limited number of spaces; a semaphore can represent the number of available spaces. Threads trying to enter the parking lot (access the resource) decrement the semaphore count. If the count is zero, they must wait. When a thread leaves the parking lot, it increments the counter, making a space available. This is useful for managing resource pools efficiently, allowing a flexible number of concurrent users while preventing overutilization. The correct use of semaphores, however, demands considerable care, as incorrect implementation can lead to deadlock.

Condition variables provide another layer of sophistication, particularly useful for coordinating threads based on specific conditions. They act as signaling mechanisms, allowing threads to wait for a particular state before proceeding. Think of a scenario where one thread produces data, and another consumes it. A condition variable can allow the consumer thread to wait until the producer thread has created data, ensuring data availability before processing. The producer, in turn, signals the condition variable when new data is ready. This avoids busy waiting, where threads repeatedly check for conditions, needlessly consuming resources. Mastering condition variables significantly increases the sophistication and efficiency of your concurrent programs.

Deadlocks represent a significant danger in concurrent programming, a situation where two or more threads block indefinitely, waiting for each other to release resources. It's a classic case of a chicken-and-egg problem. Imagine two threads, each holding one resource needed by the other. Neither can proceed, leading to a standstill. Careful design and implementation are crucial to avoid deadlocks. Techniques such as resource ordering (always acquiring resources in the same order) and timeout mechanisms (setting limits on how long a thread waits for a resource) can significantly mitigate this risk. Thorough testing and careful analysis are vital for identifying and eliminating potential deadlock scenarios.

Monitoring and debugging concurrent programs are considerably more complex than dealing with single-

threaded applications. Standard debugging tools often fall short, making it difficult to track the flow of execution across multiple threads. Specialized tools and techniques are usually required, which may involve tracing thread execution, analyzing resource usage, and carefully studying program logs to identify subtle timing issues. Such tools frequently offer visualization capabilities, which aid in understanding complex thread interactions. Proactive design that prioritizes clarity and modularity significantly eases the debugging process.
Effective synchronization demands a profound understanding of concurrent programming concepts. It's not enough just to understand the mechanics of mutexes, semaphores, and condition variables; you also need to develop a deep intuition for how threads interact. Practice is key, working on small projects that progressively introduce synchronization challenges, helps build the practical skills necessary to create efficient and robust concurrent systems. Remember: the complexities inherent in concurrent programming often demand a gradual, iterative learning approach, with practice and experimentation forming the cornerstone of mastery. Each successful synchronization mechanism implemented becomes a valuable lesson, enhancing your proficiency. The ability to effectively synchronize your threads significantly expands the realm of applications you can create, paving the way for creating sophisticated, high-performance applications.

6.3 Deadlocks

Deadlocks represent a significant challenge in concurrent programming. They occur when two or more processes are blocked indefinitely, waiting for each other to release resources that they need. Imagine a scenario with two trains approaching a single-track section; if each train waits for the other to proceed, neither will ever move. This simple analogy illustrates the core problem of deadlocks—a circular dependency where no process can progress. The consequences can range from minor performance hiccups to complete system crashes, making deadlock prevention and detection crucial skills for any concurrent programmer. Understanding the conditions that lead to deadlocks is the first step towards effective management.

Four necessary conditions must simultaneously exist for a deadlock to occur. These are mutual exclusion, hold and wait, no preemption, and circular wait. Mutual exclusion simply means that only one process can access a particular resource at a time. Hold and wait requires that a process holds at least one resource while waiting to acquire additional resources currently held by other processes. No preemption means that a resource can only be released voluntarily by the process holding it; it cannot be forcibly taken away. Finally, circular wait involves a chain of processes where each process is waiting for a resource held by the next process in the chain, forming a closed loop. The presence of all four conditions guarantees a deadlock; however, the absence of even one prevents it.

Identifying potential deadlocks within your code requires careful analysis. Begin by creating a resource allocation graph. This visual representation shows processes as nodes and resources as edges, with directed edges indicating resource requests and holdings. A cycle in this graph signifies a potential deadlock. While this approach works well for small systems, larger, more complex systems necessitate more sophisticated techniques. Static analysis tools can automatically detect potential deadlocks during the compilation or code review phase. Dynamic analysis, on the other hand, monitors the system's runtime behavior to identify deadlocks as they occur. However, these tools aren't foolproof; they may miss subtle deadlocks or produce false positives.

Deadlock prevention focuses on eliminating one or more of the four necessary conditions. Mutual exclusion is often unavoidable, especially when dealing with shared resources like files or network connections. However, we can address hold and wait by requiring processes to request all their required resources at once. This prevents a situation where a process holds resources while waiting for others, breaking the circular dependency. No preemption is usually difficult to enforce directly, though techniques like resource preallocation can sometimes mimic its effects. Finally, breaking the circular wait condition often involves imposing an ordering on resource requests; this ensures processes don't request resources in a way that creates a cycle.

This strategy may involve assigning a unique number to each resource and ensuring processes always request resources in ascending order.

Deadlock avoidance strategies aim to prevent deadlocks dynamically, without necessarily removing the four necessary conditions. These strategies utilize algorithms that analyze the current state of resource allocation to determine whether granting a resource request could lead to a deadlock. Banker's algorithm is a classical example of such a strategy; it requires processes to declare their maximum resource needs beforehand and the system maintains enough resources to ensure no deadlocks arise. However, these algorithms can be complex and impose overhead. Their suitability depends on the specific application and the trade-off between safety and performance.

Recovery from a deadlock often involves breaking the circular wait. External intervention is usually required. One approach is to terminate one or more of the processes involved in the deadlock. The choice of which process to terminate often involves prioritizing processes based on factors such as their importance, their progress, and the resources they are holding. Another approach is to preempt resources, forcibly taking resources away from a process and allocating them to another. However, preemption can be problematic; it might require careful consideration of data consistency and potential loss of work. It is crucial to ensure your chosen recovery mechanism minimizes disruption and data loss. Careful logging and monitoring practices are vital to facilitate efficient recovery and minimize disruption if a deadlock occurs. Regular testing with realistic workloads helps to identify and address potential issues, increasing resilience. A robust recovery strategy helps maintain the system's stability and integrity.

7 Advanced Programming Concepts

Metaprogramming empowers developers to write code that generates or modifies other code. Imagine a scenario where you need to create hundreds of similar classes, each with slight variations. Instead of manually writing each one, you could use metaprogramming techniques to generate the code automatically, saving time and reducing errors. This often involves using macros or code generation tools within the programming language itself, manipulating the Abstract Syntax Tree (AST) directly, or employing code generation libraries. Mastering this requires a deep understanding of the underlying language structure, as you're essentially writing code that writes code. The power lies in its ability to automate repetitive tasks and create highly adaptable software. Consider the implications for database schema generation: instead of manually writing SQL scripts for every table, a metaprogram could generate them from a concise description. The level of abstraction makes the process much cleaner and less error-prone. Similarly, creating boilerplate code for user interfaces can be automated, significantly speeding up development.

Reflection, a related yet distinct concept, allows a program to examine its own structure and behavior at runtime. This introspection capability enables dynamic adaptation and self-modification. Think of a plugin architecture where a program loads and interacts with external modules at runtime. Reflection provides the mechanism to discover and interact with these modules dynamically, extending the program's functionality without recompilation. Debugging tools also heavily rely on reflection to inspect variables and call stacks during runtime. This 'self-awareness' is crucial for creating flexible systems capable of responding to changing conditions. For instance, a game engine might use reflection to load and manage game assets, dynamically adapting to different game worlds or characters. Furthermore, configuration systems often leverage reflection to read settings from external files and dynamically adjust program behavior. The possibilities are vast, enabling considerable flexibility in software development.

Aspect-Oriented Programming (AOP) tackles the challenge of cross-cutting concerns—code that affects multiple parts of a program, such as logging, security, or transaction management. Imagine implementing logging in a traditional object-oriented program; you'd have to add logging statements to every method that requires logging. AOP offers a more elegant solution by separating these concerns into independent "aspects." These aspects weave into the main program's flow at designated points, without cluttering the core logic. This enhances modularity, reducing code duplication and improving maintainability. Consider a security aspect that enforces authentication and authorization checks; this aspect could be applied to multiple parts of an application without altering their core functionality. This approach leads to cleaner, more focused code, ultimately making it easier to understand and maintain, especially in large projects. A robust logging system, for example, can be implemented as a separate aspect, enhancing debugging capabilities without bloating the application's main logic. The benefit is a marked improvement in the clarity and maintainability of your code base.

By combining metaprogramming's code generation capabilities, reflection's introspection power, and AOP's modular separation of concerns, developers gain powerful tools for creating highly adaptable, maintainable, and efficient software systems. These advanced techniques represent a leap forward in software engineering, providing elegant solutions to complex programming challenges. The ability to manipulate code as data opens new possibilities for generating custom solutions and adapting to unforeseen situations. Mastering these concepts unlocks a new level of software development proficiency, allowing for the creation of sophisticated, flexible applications capable of evolving and adapting to changing needs. The key lies in understanding the fundamental principles of each technique and carefully applying them to specific problems. A thorough grasp of these advanced concepts distinguishes a skilled programmer from an expert, bridging the gap between proficiency and mastery in the field of software engineering.

7.1 Metaprogramming

Metaprogramming empowers programmers to write code that writes code. It's a powerful technique, allowing you to generate, modify, or analyze program code at compile time or runtime. Imagine creating a program that automatically generates boilerplate code for you, significantly reducing development time and effort. This isn't science fiction; it's a reality made possible by metaprogramming. Consider the repetitive task of creating getters and setters for numerous class members in object-oriented programming; metaprogramming can automate this process entirely.

Think of macros in C or C++. These preprocessor directives enable you to define shortcodes that expand into larger code blocks during compilation. This substitution happens before the actual compilation, enabling code generation based on patterns or rules. For example, a simple macro might replace a repeated calculation with a single call, optimizing your code for both readability and efficiency. This seemingly small act is a fundamental aspect of metaprogramming—it's about manipulating code itself to enhance the development process.

However, metaprogramming is not merely about automating simple tasks. Its application stretches to complex scenarios like code generation for domain-specific languages (DSLs). DSLs are custom-designed languages tailored to a specific problem domain, and metaprogramming allows you to create tools that translate the higher-level DSL into the target language, simplifying complex tasks. This reduces development time and allows non-programmers to interact with software using a more intuitive interface. The development of compilers and interpreters frequently leverages the concept; creating a compiler that transforms a high-level language into machine code is a sophisticated example of metaprogramming.

Templates in C++ exemplify this higher level of control. Templates allow you to write generic code that can work with various data types without needing separate implementations for each type. The compiler generates specific code for each type used, making it a form of compile-time metaprogramming. Consider the impact this has on reducing redundancy; instead of writing the same algorithm multiple times for integers, floats, and strings, a single template covers all of these. This is crucial for maintainability – fixing a bug in a template fixes it across all its instances.

Runtime metaprogramming provides yet another layer of flexibility. This involves manipulating code during the program's execution. Reflection, a key component of runtime metaprogramming, allows your code to examine and modify its own structure. Imagine a program that can inspect its own classes and methods at runtime, adapting its behavior based on the available resources or user input. This level of dynamism is especially valuable in situations requiring adaptability and dynamic code modification. A compelling example could be an application that dynamically loads plugins or modules based on user needs, adapting its functionality without recompilation.

But metaprogramming, like any powerful tool, necessitates caution. Overuse can lead to less readable and maintainable code. The generated code may become complex, potentially obscuring the original intent and hindering debugging. Furthermore, extensive use of metaprogramming can make it harder for others to comprehend and maintain your code, increasing development costs in the long run. Finding the right balance between the power of automation and the need for clear, easily maintainable code is crucial for successfully employing metaprogramming.

Finally, the choice between compile-time and runtime metaprogramming depends heavily on the specific needs of your project. Compile-time metaprogramming offers performance benefits because the generated code is integrated directly into the final executable, while runtime metaprogramming provides flexibility but can incur a performance overhead. For example, if optimization is paramount and the program structure is known beforehand, compile-time is preferable. However, if the program needs to adapt dynamically based on external factors, runtime metaprogramming would be more suitable. Understanding these trade-offs is paramount in making informed choices. The thoughtful application of metaprogramming can significantly enhance the development process, but a careful approach is needed to harness its potential while mitigating its risks. It's a powerful ally when wielded judiciously.

7.2 Reflection

Reflection allows a program to examine its own structure and behavior at runtime. This powerful capability opens doors to dynamic code generation, adaptable systems, and sophisticated debugging tools. Imagine a program that can analyze its own code, identify inefficiencies, and even rewrite itself to optimize performance—that's the potential of reflection.

Unlike static compilation, where the program's structure is fixed before execution, reflection provides a dynamic perspective. The program becomes self-aware, able to inspect its classes, methods, fields, and even modify them on the fly. This dynamic adaptability is invaluable for frameworks and libraries designed to handle diverse data structures or adapt to changing runtime environments. Think of a web server that dynamically adjusts its resource allocation based on incoming requests – reflection underpins this kind of intelligent behavior.

Consider the task of serializing an object into a format like JSON. Without reflection, you'd need to explicitly specify each field to include in the serialization process, which is tedious and error-prone. Reflection, however, enables automatic serialization, simply by introspecting the object's structure and extracting the necessary data. This drastically simplifies development, improving efficiency and maintainability. This feature is vital for frameworks, where the exact structure of the data being processed is often unknown beforehand.

Furthermore, reflection plays a crucial role in debugging and runtime analysis. Debugging tools can leverage reflection to inspect the program's state, examine variable values, and even step through execution flow at runtime. The ability to examine the internal workings of a running program is invaluable for identifying and resolving bugs. Modern IDEs make use of reflection behind the scenes to provide robust debugging features. This makes tracking down errors far easier.

However, the flexibility of reflection comes with caveats. Excessive use can lead to code that's difficult to understand, maintain, and debug. The dynamic nature of reflection can make the runtime behavior less predictable than statically compiled code. The performance overhead associated with runtime introspection should also be considered. It's crucial to use reflection judiciously, only when its benefits outweigh the potential drawbacks. Blindly employing reflection without careful consideration can lead to a performance bottleneck.

A practical example is a dependency injection framework. These frameworks utilize reflection to automatically instantiate and wire together various components within an application. By introspecting the dependencies of a class, the framework can locate and inject the necessary objects, simplifying the configuration and streamlining the development process. The ability to dynamically resolve these relationships is key to the framework's flexibility, allowing it to adapt to changing requirements. This simplifies development and promotes maintainability.

Metaprogramming builds upon reflection, leveraging the ability to inspect and modify code at runtime. While reflection allows a program to examine its own structure, metaprogramming enables programmatic manipulation of that structure. For instance, a metaprogramming technique could automatically generate code based on a set of rules or generate code from annotations in the original code. Metaprogramming significantly simplifies repetitive tasks and allows for more concise code, reducing errors and making maintenance easier.

Understanding reflection is critical for advanced programming. It's not just a niche technique. Reflection offers a level of adaptability and dynamism that's increasingly important in today's complex software environments. Mastering its principles, however, requires careful consideration and a balance between its empowering capabilities and potential drawbacks. Knowing how to use it effectively differentiates experienced programmers from beginners.

One common application of reflection lies in building dynamic proxies. These proxies intercept method calls to a target object and modify or extend the behavior. Security frameworks, for instance, frequently employ proxies to add authentication or authorization checks to method calls, without modifying the original code.

This ability to add functionality transparently without altering existing codebases significantly simplifies development and maintenance.
Reflection is also essential in building flexible configuration systems. These systems can read configuration data from various sources, dynamically map the data to program objects, and update the program's state accordingly. The runtime adaptation capabilities offered by reflection are crucial in creating systems that can seamlessly adapt to changing environments without requiring manual code adjustments. This allows for easier updating of configurations and reduces the number of updates.
Finally, let's not forget the debugging prowess reflection unlocks. Debuggers can utilize reflection to provide significantly richer diagnostics information by examining the internal state of objects and variables at various points in the execution, allowing developers to more efficiently identify and fix errors. The power to inspect a running program's internal workings is a priceless tool for programmers. It's far more efficient than traditional debugging techniques.

7.3 Aspect-Oriented Programming

Aspect-Oriented Programming (AOP) offers a powerful approach to modularizing concerns that cut across multiple parts of a program. Imagine trying to track logging information across an entire application. In traditional object-oriented programming (OOP), you'd likely end up scattering logging statements throughout your codebase, leading to a messy, hard-to-maintain system. AOP provides a cleaner solution.

Instead of embedding logging directly within your core business logic, AOP allows you to define aspects—separate modules that encapsulate cross-cutting concerns such as logging, security, and transaction management. These aspects are then woven into the application at specific points, called join points, without modifying the core application code itself. This separation significantly improves code clarity, maintainability, and reusability.

Consider a simple e-commerce application. Several components might handle user authentication: login forms, product pages, shopping carts, and order processing. With traditional methods, authentication logic would be scattered across these modules. Using AOP, authentication becomes an aspect—a reusable module containing all the security checks. This aspect would be woven into the various components at appropriate join points, such as before accessing protected resources. This centralizes security logic, making it easier to update, maintain, and audit.

AOP leverages several key concepts. Join points represent specific points in the program's execution where an aspect can be applied. Pointcuts define precisely which join points an aspect should affect. Advice is the code that the aspect executes at a join point. Advice can be executed before, after, or around a join point, providing flexibility to control program behavior. Weaving is the process of combining aspects with the base application code, creating a final executable program.

The benefits of adopting AOP extend beyond improved modularity. It enhances maintainability by centralizing cross-cutting concerns, making updates and bug fixes simpler. The reduced code duplication resulting from AOP also improves code consistency, reducing the likelihood of errors. Furthermore, AOP promotes a cleaner separation of concerns, resulting in more understandable and manageable codebases, especially in large or complex projects. This leads to improved developer productivity and faster development cycles.

However, AOP isn't a silver bullet. Overuse can lead to a decrease in code readability if aspects become overly complex or numerous. Debugging can also be more challenging in AOP applications, requiring a deeper understanding of the aspect's interactions with the core application. The initial learning curve associated with AOP concepts like join points, pointcuts, and advice can present a barrier for entry for some developers.

Languages like Java and AspectJ provide robust support for AOP. AspectJ, in particular, is a powerful extension of Java that directly integrates AOP features. Other languages offer AOP capabilities through libraries or frameworks. Choosing the right tool depends on your programming language, project requirements, and team expertise. Remember, AOP should be applied strategically to address genuinely cross-cutting concerns; it's not appropriate for every situation. Effective AOP implementation depends on careful planning and thoughtful

design.
Learning AOP requires a shift in perspective from traditional OOP paradigms. While OOP focuses on organizing code around objects and their interactions, AOP expands the focus to encompass aspects that weave across these objects, providing a more holistic perspective on modular design. Many resources exist to help navigate this shift, including online tutorials, courses, and books that provide practical guidance and real-world examples. Starting with small, well-defined aspects is key to successful AOP implementation. Begin by identifying clear cross-cutting concerns in your existing projects. Gradually integrate AOP principles into your development workflow. Remember, the goal is to enhance code quality, maintainability, and overall development efficiency, not to complicate matters unnecessarily. A well-applied aspect should enhance code clarity; poorly designed aspects hinder it. Careful consideration of design and strategic implementation is paramount.

www.ingramcontent.com/pod-product-compliance
Lightning Source LLC
LaVergne TN
LVHW020236210926
848481LV00016B/2344
* 9 7 9 8 3 4 9 2 3 3 0 1 2 *